ABOVE: THE GATEWAY ARCH in St Louis, Missouri, is based on a catenary curve (see pp. 24 – 25): $f(x) = b \cosh(x)$. When b is positive, the curve has a minimum, describing the shape of a hanging chain. When b is negative the curve is inverted and has a peak, like in this image. Illustration by Steve Scott for the 59 Parks project.

US edition © Wooden Books Ltd 2024
Published by Wooden Books LLC,
San Rafael, California

First published in the UK in 2023
by Wooden Books Ltd, Glastonbury, UK

Library of Congress Cataloging-in-Publication Data
Linton, O.
Mathematical Functions

Library of Congress Cataloging-in-Publication
Data has been applied for

ISBN-10: 1-952178-38-x
ISBN-13: 978-1-952178-38-2

Designed and typeset in Glastonbury, UK

Printed in India on FSC® certified papers by
Quarterfold Printabilities Pvt. Ltd.

MATHEMATICAL
FUNCTIONS
A VISUAL GUIDE

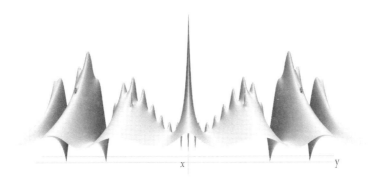

Oliver Linton

with graphs by Tihana Šare

Other books in this series by Oliver Linton include:
Fractals – On the Edge of Chaos
Numbers – To Infinity and Beyond.
A further selection of articles by the author on a
range of topics can be found at www.jolinton.co.uk.

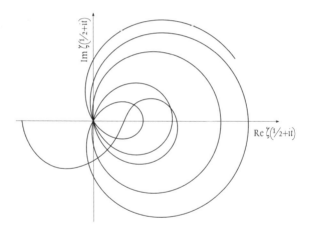

ABOVE: *A parametric plot of the Riemann zeta function along the critical line (see p. 54).*
The points where the line passes through the origin represent the famous 'zeros' which lie at the
heart of one of the most important unsolved mysteries in Mathematics. Four of these 'zeros' can
also be seen in the illustration on the title page. TITLE PAGE: *The modulus of the Riemann zeta*
function in the region $R(z) > 0$. *The line* $R(z) = \frac{1}{2}$ *is indicated. The vertical dimension has been*
greatly reduced for this 3D visualization. The four down spikes mark the four mysterious zeros.

CONTENTS

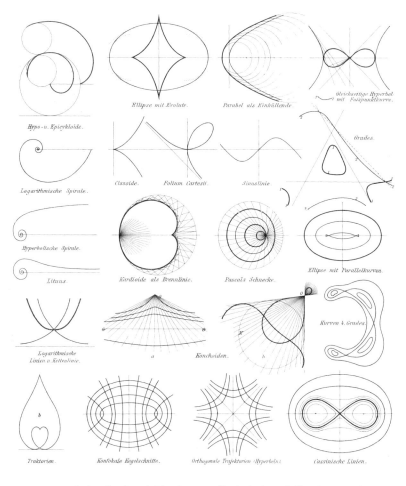

Hypo- u. Epicykloide.

Ellipse mit Evolute.

Parabel als Einhüllende.

Gleichseitige Hyperbel mit Fußpunktkurve.

Logarithmische Spirale.

Cissoide.

Folium Cartesii.

Sinuslinie.

Grades.

Hyperbolische Spirale.

Lituus.

Kardioide als Brennlinie.

Pascals Schnecke.

Ellipse mit Parallelkurven.

Logarithmische Linien u. Kettenlinie.

a

Konchoiden.

b

Kurven 4. Grades.

Traktorien.

Konfokale Kegelschnitte.

Orthogonale Trajektorien (Hyperbeln).

Cassinische Linien.

ABOVE: Early plots of mathematical functions. Some (like the simple parabola) are best graphed using Cartesian (x/y) coordinates. Others (like the cardioid) are best graphed using polar coordinates. Some (like the ellipse) can be described in either coordinate system equally well.

INTRODUCTION

T HERE IS PROBABLY no single more important idea in all of Mathematics than the idea of a *function*, whether the formula for calculating your body mass index, BMI = mass / height^2, or a table of values showing the number of days in each month. It was Descartes in the seventeenth century who first discovered how to turn a mathematical equation into a picture, and by observing how the plot changes with different numerical values, how to appreciate the beauty and subtleties concealed within the often complicated-looking formulae.

A mathematical function is defined as a procedure by which an input value (taken from a possibly restricted set called the *domain* of the function) is paired with a unique output value. By plotting the input value on the x-axis and the output value on the y-axis of a piece of squared paper you obtain a picture of the function (a *graph*) which makes the salient features of the function immediately apparent. If the input value is an *angle* rather than a simple number, it may be appropriate to plot the function in polar coordinates using graph paper with concentric circles around the origin instead of squares.

Shown on the left is a plate of famous mathematical curves from a late nineteenth century textbook. The caption explains how they can be graphed in different ways. Familiarising yourself with functions will enable you to look at plants, weather charts and architectural forms in new ways, and get you thinking about how the bends and surfaces of natural and man-made objects can be approximated and represented by mathematical equations. Your journey into the exciting world of mathematical functions begins here!

BASIC FUNCTIONS
from algebra to calculus

Descartes' discovery that an algebraic equation could be plotted as a graph turned many difficult geometrical problems into easy algebraic ones. For example, suppose a farmer wants to find the biggest rectangular field he can enclose with 100m of fencing placed along an existing wall. He could do it by drawing a range of scale diagrams and calculating their areas to see which was largest, but instead, if he takes the length of the fence parallel to the wall as x, then the length of the fencing perpendicular to the fence must be $\frac{1}{2}(100 - x)$, so the area A must be $x \times \frac{1}{2}(100 - x)$, so $A = 50x - \frac{1}{2}x^2$. Now all he has to do is plot a graph of this function (it turns out to be an inverted parabola, *below*) and see at what value of x it reaches its maximum value.

Descartes' discovery opened up a whole new field of enquiry because it suddenly became possible to visualise and study how things change with time. Galileo had discovered in 1632 that falling objects accelerated at a constant rate and that the trajectory of an ideal cannon ball was a parabola, but what did this discovery mean as regards the motion of the planets or the behaviour of hurricanes? Barely fifty years later, Leibnitz and Newton had discovered the key to unlocking the full potential of the idea of a function—calculus (*p.32*).

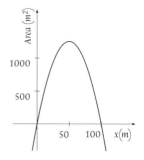

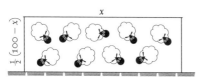

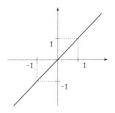

Linear function

$f(x) = x$

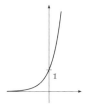

Exponential function

$f(x) = 10^x$

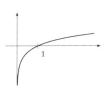

Logarithmic function

$f(x) = \log x$

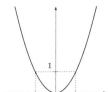

Square (parabolic) function

$f(x) = x^2$

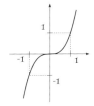

Cubic function

$f(x) = x^3$

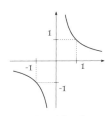

Reciprocal function

$f(x) = {}^1\!/x$

ABOVE: Some simple functions. Some of these (like the parabola) can be reflected left/right about the vertical axis and are said to be 'even' (as they only contain only even powers of x). Others (the linear, cubic and reciprocal functions) have rotational symmetry about the origin and are said to be 'odd' (as they contain only odd powers of x).

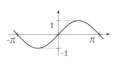

Sine function

$f(x) = \sin x$

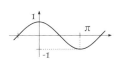

Cosine function

$f(x) = \cos x$

Tangent function

$f(x) = \tan x$

ABOVE: The trigonometric functions found in a right-angled triangle are also essential in describing oscillating systems such as those found in acoustics and radio. They are periodic functions, meaning that for a period $T > 0$ the function has a repeating value, $f(x + T) = f(x)$.

LINEAR FUNCTIONS
getting things in proportion

To measure temperature, Americans use the Fahrenheit scale, while Europeans (and much of the rest of the world) use the Celsius scale. In the eighteenth century, Fahrenheit's 0° was defined as the lowest temperature which could be achieved using a mixture of ice and salt; body temperature was defined as 90°. Later it was realised that neither of these 'fixed' points were very stable and the scale was redefined so that the freezing point of water was exactly 32° and the boiling point of water was 212° (putting body temperature at 98.2°).

There is a simple procedure for converting from Fahrenheit to Celsius. First subtract 32, then multiply by 5 and divide by 9. This can be expressed as an equation:

$$°C = (°F - 32) × 5/9.$$

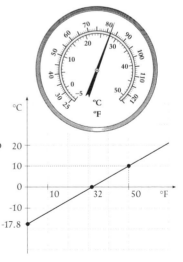

A graph of this function is shown (*right*). It takes the form of a straight line which cuts the *x*-axis at 32° (because 32°F = 0°C), and the *y*-axis at –17.8 (because 0°F = –17.8°C), and has a *gradient* (or slope) of 5/9 (because a rise of 9°F is equivalent to a rise of 5°C).

For obvious reasons, a function like this is called a *linear function*. Importantly, linear functions have no squared or cubed terms—the highest degree of *x* is 1.

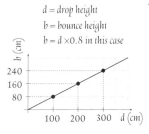

d = drop height
b = bounce height
b = d × 0.8 in this case

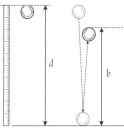

LEFT: A bouncing ball. The height of each bounce, b, is a linear function with the drop height, d, as its variable: b = d × c, where c is a constant representing the bounciness of the ball.

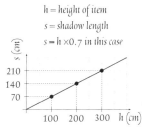

h = height of item
s = shadow length
s = h × 0.7 in this case

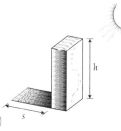

LEFT: The length of a shadow, s, cast by an object from a fixed light source, is a linear function with the object height, h, as its variable: s = h × c, where c is a constant.

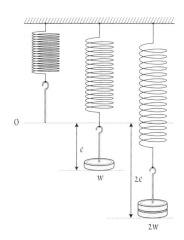

LEFT: Almost all physical laws involve functions in one way or another. E.g. Hooke's Law states that the extension of a spring, e, is proportional to the weight, w, hung on it (up to a point!). Thus, doubling the weight doubles the extension. A graph of the extension against weight is a straight line passing through the origin (0, 0). A graph of the total length of the spring, l, against weight has the same gradient, but does not start at the origin.

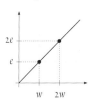

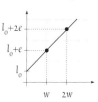

POWER FUNCTIONS
shapes and energies

Some quantities, like kinetic energy, obey what is known as a power law. A car travelling at 60 mph has four times as much kinetic energy as a car travelling at 30 mph and will take *four* times the distance to stop. This is because kinetic energy is proportional to the square of the velocity, $\text{KE} = \frac{1}{2}mv^2$. A graph of kinetic energy against speed is therefore a curve and, since a car travelling backwards has the same kinetic energy as a car travelling forwards, it is symmetrical about the *y*-axis. The graph has the shape of a *parabola* (*opposite top*).

The volume of a jug is proportional to the *cube* of its size so a jug twice the size (in all directions) will contain *eight* times as much water as a standard jug. It makes no sense to talk of jugs with negative size but, mathematically, there is nothing wrong with cubing a negative number (the cube of $-x$ is simply equal to $-x^3$). This has an important implication as regards the shape of the curve.

The Sun has a surface temperature of 6000°C and warms the Earth at a rate of about 3 kW over every square metre. Now the rate at which a body gives off heat depends on the fourth power of its surface temperature. So if the Sun was twice as hot (i.e. if it was a white dwarf) Earth would receive energy at 16 times the rate and our oceans would boil.

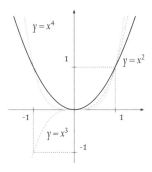

The higher the power of the function the flatter the curve is at the start and the steeper it is at the end.

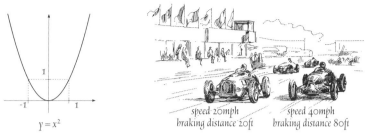

$y = x^2$

speed 20mph
braking distance 20ft

speed 40mph
braking distance 80ft

ABOVE: The kinetic energy of a car is proportional the square of its speed. Graphs of the form $y = cx^2$ display the distinctive curve of a parabola. When c is negative, the parabola will be inverted vertically.

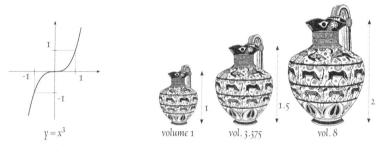

$y = x^3$

volume 1

vol. 3.375

vol. 8

ABOVE: The graph of $y = x^3$ is anti-symmetric. The volume of a regular three-dimensional object is proportional to the cube of its size, so doubling the size of a jug means it holds eight times more wine.

$y = x^4$

Sun, temperature 1
radiated energy 1

white dwarf, temperature 2
radiated energy 16

ABOVE: The graph of $y = x^4$ is bowl shaped. Examples in nature are rare, however, the energy radiated from a fire is proportional to the fourth power of its (absolute) temperature.

POLYNOMIAL FUNCTIONS
just wiggly lines

The most familiar mathematical functions are those defined by a polynomial equation of the form

$$f(x) = y = a_n x^n + a_{n-1} x^{x-1} + \cdots + a_1 x + a_0$$

The *domain* of this function is the set of real numbers and its *co-domain* (output) is the same, which is why it is possible and convenient to draw a graph of the function using Cartesian coordinates.

The powers of x (i.e. x^1, x^2, x^3, x^4, x^5, x^6 ... etc) are called *exponents* and the highest exponent of x determines the *order* or *degree* of the function. A glance at the examples (*opposite*) will reveal that the higher the order of the function, the more wiggles it can (potentially) have. Indeed, it is clear that a function of order n can intersect a straight line in no more than n places. (Some functions like $f(x) = x^4$ are *degenerate* in that there is no line that cuts the graph in four places.)

A good way to predict the shape of a polynomial function is to consider the limiting cases when x is very large or very small. Supposing, for example, that we want to sketch the function defined by the equation: $$f(x) = y = x^5 - x^2 + 1$$

If x is very large, we can ignore the x^2 term and the 1 in comparison with the x^5 term and since this term has an odd power, the graph will shoot off to $+\infty$ in the positive direction and $-\infty$ in the negative direction. But when x is small, the x^5 term can be ignored, so here the function will look like an inverted parabola peaking at $+1$ when $x = 0$. The absence of any cubic or quartic terms suggests that this function is partially degenerate and there will be no more wiggles in it.

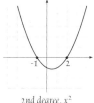

2nd degree, x^2

$$f(x) = (x+1)(x-2)$$
$$= x^2 - x - 2$$

3rd degree, x^3

$$f(x) = \tfrac{1}{4}(x+4)(x+1)(x-2)$$
$$= \tfrac{1}{4}(x^3 + 3x^2 - 6x - 8)$$

4th degree, x^4

$$f(x) = \tfrac{1}{14}(x+4)(x+1)(x-1)(x-3)$$
$$= \tfrac{1}{14}(x^4 + x^3 - 13x^2 - x + 12)$$

5th degree, x^5

$$f(x) = \tfrac{1}{22}(x+4)(x+2)(x+1)$$
$$(x-1)(x-3)$$

6th degree, x^6

$$f(x) = \tfrac{1}{4}(x+3)(x+2)(x+1)$$
$$x(x-1)(x-2)$$

7th degree, x^7

$$f(x) = \tfrac{1}{17}(x+3)(x+2)(x+1)$$
$$x(x-1)(x-2)(x-3)$$

ABOVE: The greater the polynomial degree, the greater the wiggle. The polynomial functions are shown factorised, so zeros can be read immediately; e.g. for $f(x) = (x+1)(x-2)$, when either of the brackets is zero, $f(x)=0$, so the zeros for that 2nd degree function are $x = -1$ and $x = 2$.

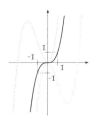

$y = x^3 + cx^2$

Different values of c $(-3, 0, 3)$ vary the amount of wiggle. All can be cut by a straight line in 3 places.

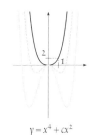

$y = x^4 + cx^2$

When c is negative the graph has the characteristic shape of a quartic curve $(c = -7, -3, 1)$.

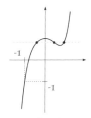

$y = x^5 - x^2 + 1$

This curve can only be cut in three places even though it is a quintic.

Reciprocal Functions
zeros shoot off to infinity

Reciprocal functions (more correctly called 'rational functions') have the form:

$$f(x) = \frac{a_n(x)}{a_m(x)}$$

where $a_n(x)$ and $a_m(x)$ are polynomials of order n and m respectively.

The function will have order n or $m+1$, whichever is the greater. If there are values of x for which $a_m(x) = 0$ then the function will shoot off to ∞ at these points (known as *singularities*), whereas if $a_m(x)$ is never equal to zero, then the function will contain no singularities. An important function in this latter category is $y = 1/(x^2 + c^2)$ (*opposite*).

The simplest reciprocal function is $f(x) = 1/x$, a rectangular hyperbola shooting off to ∞ at $x = 0$. If you approach from the positive direction $f(x)$ tends to $+\infty$; but if you approach from the negative direction $f(x)$ tends to $-\infty$. Mathematicians get over this by saying that $1/x$ is *undefined* at $x = 0$. Alternatively, we can say that the point $x = 0$ is specifically excluded from the domain of the function.

When you tune an old radio (*e.g. the 1940s spy radio below*), the response of the tuned circuit responds according to an equation similar to $1/(x^2 + c^2)$ where x is the difference between the frequency of the tuned circuit and the radio signal, and c is the 'Q factor' of the circuit (how focused it is). Circuits with a small Q are good at distinguishing between nearby stations but require precise tuning.

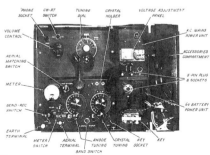

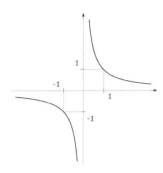

$$y = {}^1/x$$

This is a rectangular hyperbola. It has order 2, as a straight line can cut it in at most 2 places.

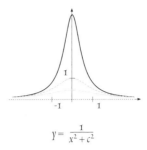

$$y = \frac{1}{x^2 + c^2}$$

Different values of c produce different degrees of 'humpiness'. All these graphs can be cut by a straight line in three places. The graph could represent the intensity of the sound made by a car as it passes you at distance c, or the tuning range of a radio, as explained on the facing page.

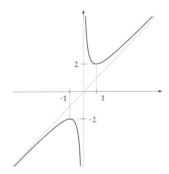

$$y = \frac{x^2 + 1}{x}$$

Some reciprocal functions have asymptotes, i.e. lines towards which the function tends as x approaches infinity. This function has a diagonal asymptote because when x is large the equation reduces to $y = x$.

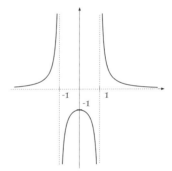

$$y = \frac{1}{x^2 - 1}$$

The denominator goes to zero at $x = 1$ or -1 so this cubic function has two singularities. Cubics are cut by a straight line in either 1 or 3 places, but horizontal lines cut the curve in either 2 places or none (as with y constant it becomes a quadratic).

TRIGONOMETRICAL FUNCTIONS
oscillating systems

You may recall from school that in a right-angled triangle with a hypotenuse of 1, the *sine* of x is equal to the length of the opposite side and the *cosine* of x is equal to the length of the adjacent side. While this is true, it is not really how mathematicians look at trigonometric functions. In the first place, x is not really an *angle*, it is just a number. Moreover, the functions of $\sin x$ and $\cos x$ are defined in such a way that they repeat with a cycle length of 2π. Looked at this way, the argument of the function (x) can take any value from $-\infty$ to $+\infty$ with the result being a number between -1 and $+1$. Therefore, sine and cosine are *continuous* functions.

The *tangent* function is defined as $\frac{\sin x}{\cos x}$ and looks different from $\sin x$ and $\cos x$ because of its singularities. The other trigonometrical relations, the *secant*, *cosecant* and *cotangent*, are the reciprocals of the cosine, sine and tangent, respectively.

Like the reciprocal function itself, all of these (except sine and cosine) have singularities (at multiples of $\pi/2 \pm n\pi$); but notwithstanding that, they are all perfectly good, *single-valued* functions of x.

The sine and cosine functions can be represented by any oscillatory system such as a swinging pendulum or a simple harmonic oscillator.

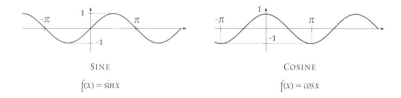

SINE

$f(x) = \sin x$

COSINE

$f(x) = \cos x$

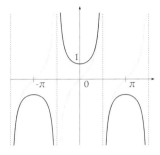

SECANT

$$f(x) = \sec x = \frac{1}{\cos x}$$

Both sec x (black) and tan x (grey) have
singularities at the same place. Where
tan x is zero, sec x has its peaks.

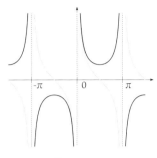

COSECANT

$$f(x) = \operatorname{cosec} x = \frac{1}{\sin x}$$

The graph of cosec x is the same as sec x but
shifted by $\pi/2$. Cosec x (black) and cot x (grey)
have singularities at the same place.

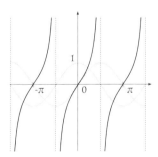

TANGENT

$$f(x) = \tan x = \frac{\sin x}{\cos x}$$

The graphs of sine and cosine are the same but
shifted (grey), while the tangent graph (black)
has singularities whenever cos x = 0.

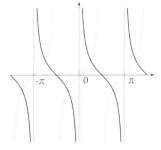

COTANGENT

$$f(x) = \cot x = \frac{1}{\tan x} = \frac{\cos x}{\sin x}$$

The cotangent function (black) has the
same shape as the tangent function
(grey) but reflected and shifted by $\pi/2$.

13

MULTI-VALUED FUNCTIONS
implicit and parametric equations

All the functions we have considered so far have been what is known as *explicit* functions: the value of the function $f(x)$ can be expressed in the form of an equation containing only the variable x.

Take, however, the equation $x^2 + y^2 = 1$. This can be turned into the function $f(x) = \sqrt{(1 - x^2)}$. Equations like this in which x and y terms are mixed up together are called *implicit* equations, and the functions they produce can be multivalued, as every positive number has two square roots. Importantly, not all implicit equations can be easily turned into explicit ones (e.g. the Folium of Descartes, $x^3 + y^3 = 3xy$, *see page 20*).

The square root function is shown below. Although it appears to be multi-valued, as a vertical line cuts the graph in two places, mathematicians instead regard this as two separate functions with different codomains, one giving the positive root, the other giving the negative root. The implicit equation for the square root function is $x - y^2 = 0$, and the multi-valued equation is $y = \sqrt{x}$.

The circle and ellipse functions (*opposite*) can also be understood as either one multi-valued function or two separate functions. Another way of defining an implicit equation is in terms of two equations, one for x and one for y, in terms of a third variable (often θ or t). Such equations are called *parametric equations* and some examples are given on the right.

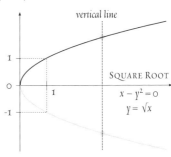

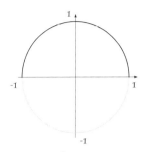

CIRCLE
implicit: $x^2 + y^2 = 1$
multivalued: $y = \sqrt{(1 - x^2)}$
parametric: $x = \cos\theta$; $y = \sin\theta$

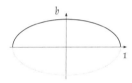

ELLIPSE
implicit: $x^2/a^2 + y^2/b^2 = 1$
multivalued: $y = b\sqrt{(1 - x^2/a^2)}$
parametric: $x = a\cos\theta$; $y = b\sin\theta$

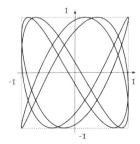

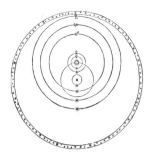

ABOVE: Tycho Brahe's model of the solar system.
All early models of the solar system held that the
planets moved in perfect circles.

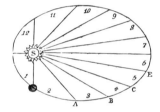

ABOVE: According to Kepler's laws, planets
orbit the Sun in ellipses, with the Sun at one
focus, sweeping out equal areas in equal time.

LISSAJOUS FIGURE
$$x = \cos(3t) \qquad y = \cos(5t + 0.2)$$

Lissajous figures are a class of parametric
functions defined by the pair of equations:
$$x = \cos(nt) \qquad y = \cos(mt + \phi)$$
where t is time elapsed, and ϕ alters the phase
relation. Pretty pictures appear when n and m
are in simple ratios, e.g. 2:3, or 3:5 as shown.

15

POLAR FUNCTIONS
spirals, fish and butterflies

In addition to using Cartesian (x, y) coordinates, further curves can be generated using *polar coordinates* (θ, r) in which a length r, called the *modulus*, is defined as a function of the *argument*, an angle θ.

The polar equation $r = a$ generates a circle of radius a, while $r = a\theta$ produces an equally-spaced *Archimedean spiral*. The spiral commonly found in nature (e.g. in sunflower heads and snail shells) is the *logarithmic spiral*, $r = ae^{\theta}$ (*both shown opposite*). Logarithmic spirals are self-similar, appearing the same when magnified to any scale.

Some polar functions employ trigonometrical functions. For example, $r = \cos(n\theta)$ generates a daisy-like curve with $2n$ petals for even (and n petals for odd) values of n. A variety of daisy shapes can be drawn by altering the initial equation (*below*). Another interesting polar function is the cardioid $r = 1 + \cos\theta$, which can be seen when a strong light is shone onto a circular cup of tea (*opposite*). Simple polar equations can be used to generate many other playful symbols.

DAISY CURVE
$r = 1 + n\cos n\theta$
$n = 4, 6, 12$

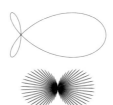

FISH/SCALLOP CURVE
$r = \cos\theta\cos 2\theta$
$r = \cos\theta\cos 64\theta$

BUTTERFLY CURVE
$r = 2\sin 2\theta - \sin n\theta$
$n = -8, 4, 6.$

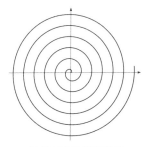

ARCHIMEDEAN SPIRAL
$$r = a\theta$$
where a is a constant, used to define the degree
of the equal spacing of the arms of the spiral

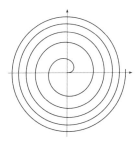

FERMAT'S SPIRAL
$$r^2 = a^2\theta$$
The 'parabolic' spiral, where a is a constant.
Encloses an equal area every turn.

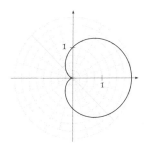

CARDIOID CURVE
$$r = 1 + \cos\theta$$
showing the polar coordinate grid used
to plot all the examples on this page.

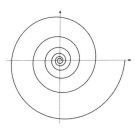

LOGARITHMIC SPIRAL
$$r = ae^{b\theta}$$
The 'exponential' or 'equiangular' spiral, where a
and b are constants and e is Euler's no. 2.71828…

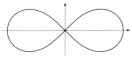

LEMNISCATE CURVE
$$r = \sqrt{\cos 2\theta}$$

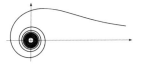

LITUUS SPIRAL
$$r = a/\sqrt{\theta}$$

ROLLING FUNCTIONS
cycloids, epicycloids and hypocycloids

The trajectory of a point on the surface of a circular wheel, radius 1, rolling along a straight line is a *cycloid* (*below*). If the wheel rolls around the outside of a circle of the same radius, the curve traced out is a *epicycloid* with one cusp, a *cardioid*. If the central circle has radius n times larger, the epicycloid will have n cusps. The parametric equations are:

$$x = (n+1)\cos\theta - \cos((n+1)\theta) \qquad y = (n+1)\sin\theta - \sin((n+1)\theta)$$

where θ is the angle through which the rolling circle has rotated. If a wheel rolls around *inside* a fixed circle n times larger, the curve traced out will be a *hypocycloid* with n cusps, given by:

$$x = (n-1)\cos\theta + \cos((n-1)\theta) \qquad y = (n-1)\sin\theta - \sin((n-1)\theta)$$

More generally, if the two circles have radii a and b (where $a > 1$), the number of cusps of an epicycloid or hypocycloid will be the numerator (top numeral) of the simplified fraction b/c ($= a/b$).

Instead of drawing the path of a point on the perimeter of the wheel, Further patterns can be obtained by drawing the path of points within the wheel (*as in a spirograph, opposite*), or on an arm attached to the wheel and extending beyond it.

CYCLOID - *a combination of uniform motion in a straight line and in a circle, of the form:* $x = r(\theta + \sin\theta)$ *and* $y = r(1 - \cos\theta)$, *where*

r is the distance of a point from the centre of the wheel. If r is greater than 1, there will be times when the point is moving backwards.

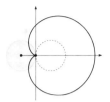

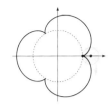

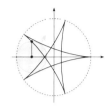

CARDIOID
The special case of an epicycloid where both circles are the same size.

EPICYCLOID
A 3-cusp pattern. The outer wheel is one third of the inner.

HYPOCYCLOID
A 5-cusp pattern. The inner wheel is two fifths of the outer.

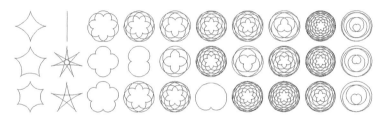

ABOVE: A selection of epicycloid and hypocycloid curves.

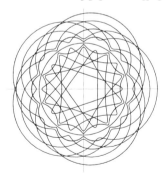

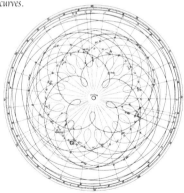

ABOVE: A family of epicycloid and hypocycloid spirograph patterns, the variations from changing the position of the pen in the wheel.

ABOVE: The paths of the planets Venus (5 cusps) and Mercury (22 cusps) around the Earth take the form of hypocycloid curves.

INVERSE FUNCTIONS
line of reflection

A function is a formula for generating y given x. The inverse of a function is a formula for generating x given y. Even if a formula is single-valued its inverse function will often turn out to be multi-valued, e.g. the inverse of the function $y = x^2$ is $x = \pm\sqrt{y}$.

It is easy to see why. The graph of the inverse of a function looks the same but with the axes swapped round (as if reflecting the graph through the $45°$ line $y = x$). Any function which is symmetrical in x and y stays the same when x and y are swapped, so is its own inverse. Obvious examples are $y = x$ and $y = 1/x$ (*below left*); $x^2 + y^2 = 1$ (a circle); and the Folium of Descartes (*below right*), etc. On the other hand, an equation such as $y = x^3 - x$ is not symmetrical in x and y; so its inverse (*the dotted curve opposite top right*) is different, cutting the y-axis in three places, clearly showing it is a multi-valued function.

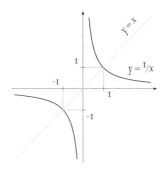

$y = x$ and $y = 1/x$ *are the only single-valued functions which are their own inverses.*

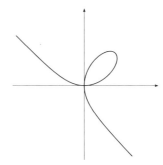

FOLIUM OF DESCARTES
$x^3 - 3xy + y^3 = 0$

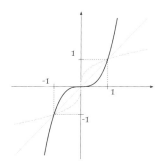

$y = x^3$

Its inverse $x = \sqrt[3]{y}$ is shown as a gray dashed line. Note that it is single-valued, as there is only one (real) cube root of a (real) number.

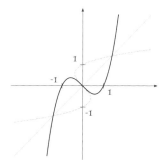

$y = x^3 - x$

Its inverse, shown as a dashed line, cuts the y-axis in three places, showing that y can take multiple values for a single value of x.

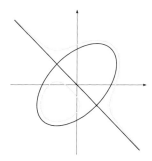

$x^3 - x - y + y^3 = 0$

This has the strange shape that it has because the left-hand side is the product of two functions: $(x + y)(x^2 - xy + y^2 - 1)$. The condition that the function should be equal to zero is satisfied if either part is zero. If the right-hand bracket is not zero the function splits in a rather surprising way.

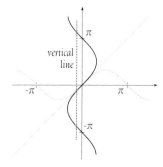

$y = \arcsin x$

The inverse of the sine function ($y = \arcsin x$ or $y = \sin^{-1} x$) returns the angle whose sine is x. Obviously x must lie between 0 and 1 and there are many angles which all have the same sine. The other inverse trig functions $\arccos x$ and $\arctan x$ are likewise multivalued.

EXPONENTIAL AND LOGARITHMIC
growth, decay & distribution

The exponential function $y = a^x$ is self-similar—if you look at the 'flat' region where x is negative and then adjust the scale of the y-axis, you will find that it looks the same as the whole graph! Its inverse is the logarithmic function $y = \log_a x$, where a is the base of the logarithm. While the domain of the exponential function is all real numbers, the logarithm function is restricted to the positive numbers.

The Gaussian function, 'bell curve', or 'normal distribution curve' is important in statistics. It models how variable quantities, e.g. the heights of 16-year-olds, often cluster around a mean value. It has a hump in the middle and two symmetrical tails which decrease rapidly to zero (*below left*). The x^2 term ensures that it is symmetrical about the mean, and the reciprocal function causes it to tend to zero as x gets further away from the mean.

Another exponential function used in statistics is the Poisson curve, which expresses the probability of a given number of random but regular events occurring in a fixed interval of time or space (*below right*).

Another important curve in this family is the exponential decay curve, $f(x) = e^{-x} \cos x$, which is used widely in physics (*lower, opposite*).

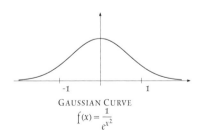

GAUSSIAN CURVE
$$f(x) = \frac{1}{e^{x^2}}$$

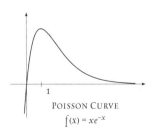

POISSON CURVE
$$f(x) = xe^{-x}$$

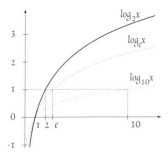

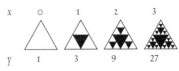

ABOVE: Sierpinski triangles. The number of white triangles y at step x increases exponentially, as $y = 3^x$ and inversely $x = \log_3 y$.
LEFT: Three different log curves.

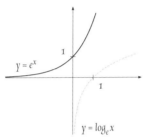

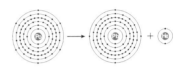

ABOVE: Radioactive decay of polonium-210 (half-life of 138 days) into lead + helium. After t days, the quantity $y(t)$ of Polonium left undecayed from some initial quantity y_0 will be: $y(t) = y_0 e^{-kt}$, where k is $\log_e 2 / 138$

NATURAL EXPONENTIALS & LOGARITHMS

ABOVE: $y = e^x$ and its inverse $y = \log_e x$. These 'natural' exponentials and logarithms employ Euler's number $e = 2.71828\ldots$ as a base, so the gradient of $y = e^x$ equals its value at every point.

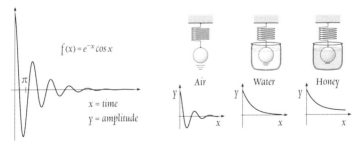

EXPONENTIAL DECAY CURVE

$f(x) = e^{-x} \cos x$, with an example of oscillators in three different mediums. When the medium is air the oscillator is under-damped and will reach the equilibrium state after some oscillations.

Hyperbolic Functions
sinh and cosh

The exponential function $f(x) = e^x$ increases without limit, while its reciprocal $f(x) = 1/e^x = e^{-x}$ decreases steadily to zero (*solid and dashed grey lines below*). The mean of these (i.e. half their sum) is an important function called the *hyperbolic cosine* or cosh x (*black line below*), while half the difference between them generates the *hyperbolic sine* or sinh x:

$$\cosh x = \frac{e^x + e^{-x}}{2} \qquad\qquad \sinh x = \frac{e^x - e^{-x}}{2}$$

It is easy to deduce from these definitions that:

$$\cosh x + \sinh x = e^x \qquad\qquad \cosh x - \sinh x = e^{-x}$$

and that therefore, by multiplying these two equations together:

$$\cosh^2 x - \sinh^2 x = 1$$

Hyperbolic functions parametrize a hyperbola the same way trigonometric functions parametrize a circle, and they crop up in many areas of mathematics and physics. The curve formed by a heavy chain hanging from two supports is the hyperbolic cosine curve, or *catenary* (*see opposite bottom*), also used in suspension bridges. Inverted, it is often used in arched structures such as the catenary arches in Casa Mila by Gaudí, and the Gateway Arch at St Louis, Missouri (*see frontispiece*).

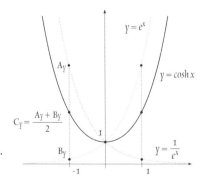

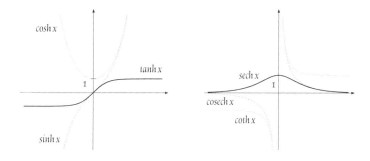

ABOVE: *The hyperbolic functions:* $y = \cosh x$; $y = \sinh x$; *and* $y = \tanh x$; *along with their reciprocals:* $y = \operatorname{sech} x = {}^{1}/\cosh x$; $y = \operatorname{cosech} x = {}^{1}/\sinh x$; *and* $y = \coth x = {}^{1}/\tanh x$.

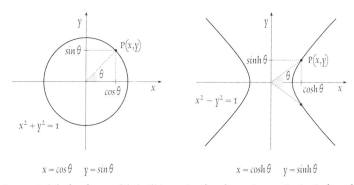

$$x = \cos\theta \quad y = \sin\theta \qquad\qquad x = \cosh\theta \quad y = \sinh\theta$$

ABOVE: *A circle* $x^2 + y^2 = 1$, *and the implicit equation* $x^2 - y^2 = 1$, *whose graph takes the form of a hyperbola. The* sinh *and* cosh *functions have the same relation to the hyperbola as* sin *and* cos *have to the circle, which is why they are known as the 'hyperbolic' functions.*

ABOVE: *A hyperbolic cosine curve, also known as a 'catenary'.*

FOURIER FUNCTIONS
analysing sounds

Put a pencil on a piece of paper and draw a continuous wiggly line from left to right. Add some axes and you have defined a function. It seems unlikely that you could write down an algebraic formula for such an arbitrary function, but in 1822 French mathematician Joseph Fourier realised that you could, using the sine function. The trick is to express the function as the sum of a series of 'harmonics' whose amplitude and phase are adjusted to achieve the desired result:

$$f(x) = a_1\sin(x + b_1) + a_2\sin(2x + b_2) + a_3\sin(3x + b_3) + \dots$$

where the *a* terms are the amplitudes and the *b* terms the phase adjustments. Although you might need an infinite number of such terms to perfectly model a function like the one you have drawn, in practice, we can often make do with just a few (*see examples opposite*).

Many digital synthesizers create unique sounds by adding sine waves. Conversely, sounds such as bird calls and voice prints are regularly analysed by breaking them down into sinusoidal components. The process of calculating the right parameters for the amplitude a_1, a_2, a_3 etc, and the phase b_1, b_2, b_3, etc, is called *Fourier analysis* and it has widespread applications in many fields of mathematics and physics.

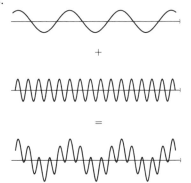

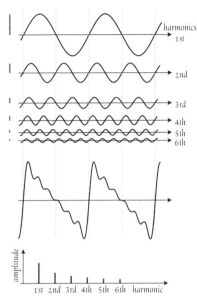

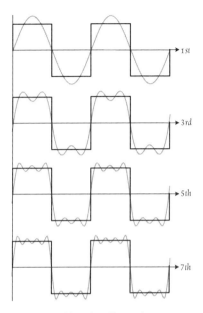

ABOVE: The first six harmonics sum to form a negative ramp sawtooth wave. As further higher harmonics are added, the sawtooth waveform becomes more perfect.

ABOVE: Odd-numbered harmonics sum to approximate a square wave. As more and more odd harmonics are added, the waveform becomes more and more square.

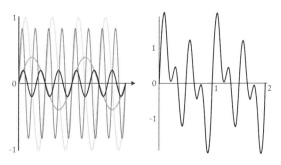

LEFT: A synthesized violin waveform, created from the first four harmonics, with the amplitudes of the 2nd and 4th harmonics greater than the amplitudes of the 1st and 3rd harmonics.

GRADIENTS
up hill and down dale

A groundbreaking discovery in mathematics was that the gradient of a function is another function, which can be deduced using a process called *differentiation*. The first *gradient/derivative* function gives the slope of the function at any given point on the graph. The general formula for differentiation of each term is:

$$f(x) = ax^n \implies f'(x) = anx^{n-1}, \text{ where } n > 0.$$

For example, the polynomial $f(x) = x^3 - x^2 - x + 0.75$ is a typical cubic function with one *maximum* and one *minimum*. The equation for its derivative function is $f'(x) = 3x^2 - 2x - 1$ (the constant 0.75 disappears). This equation can be differentiated further, giving a third function: $f''(x) = 6x - 2$. And so on, each time the order of the function reducing by one, until there is nothing left. $f'(x)$ is the first derivative of f with respect to x, $f''(x)$ is the second derivative, etc (*opposite, top*).

Uniquely, the value of the exponential function $f(x) = e^x$ is equal to its gradient at every point, so it is its own derivative $f'(x) = e^x$ (*opposite, centre*). The derivative of $f(x) = \log_e x$ is simply $1/x$ (*opposite, bottom*).

Some functions cannot be fully differentiated. For example, the function $f(x) = e^{-|x|}$ (where $|x|$ is the *absolute value* or *magnitude* of x) has a kink in it at $x = 0$, so can have no derivative at this value (*below*).

$f(x) = e^{-|x|}$

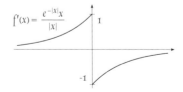

$f'(x) = \dfrac{e^{-|x|}x}{|x|}$

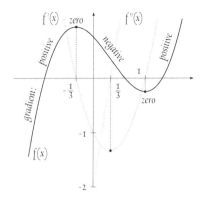

PEAKS AND TROUGHS OF FUNCTIONS

The peaks and troughs of $f(x)$ occur where its gradient, $f'(x)$, is zero. In this example, the derivative function crosses the x-axis in two places, with positive values on each side and negative values in between. Where $f'(x)$ is positive, $f(x)$ has a positive gradient, and where $f'(x)$ is negative, $f(x)$ has a negative gradient.

$$f(x) = x^3 - x^2 - x + 0.75$$
$$f'(x) = 3x^2 - 2x - 1$$
$$f''(x) = 6x - 2$$

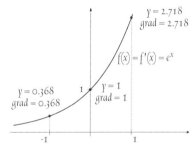

THE EXPONENTIAL FUNCTION

The exponential function $f(x) = e^x$ is equal to its gradient. No other function has this property (which determines the value of $e = 2.71828...$). At $x = -1$, $y = 1/e$ and the gradient also $= 1/e$. At $x = 0$, $y = 1$ and the gradient also $= 1$. At $x = 1$, $y = e$, and the gradient also $= e$; etc.

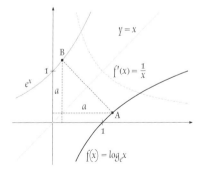

DERIVATIVE OF $\text{Log}_e(x)$

The gradient of $\log_e(x)$ is $1/x$. The log function is the inverse of the exponential function so has the same shape but reflected about the about the $45°$ line, $y = x$. Therefore, any point A on the log graph, where $x = a$, has a corresponding point B on the exponential graph, where $e^x = a$. As we have seen, the derivative of the exponential function at point B is equal to the value of the function at B. This means that the derivative of the log function at the point A is equal to $1/a$ and in general, the derivative of $\log_e(x)$ is $1/x$.

MORE GRADIENTS
differentiating trigonometrical functions

Remarkably: the gradient of the $\sin x$ function is $\cos x$; the gradient of the $\cos x$ function is $-\sin x$; the gradient of the $-\sin x$ function is $-\cos x$ and the gradient of the $-\cos x$ function is $\sin x$. So, differentiating *four times* gets you back to the original function (*shown below*).

Why is this so important? In many physical processes such as the oscillations of a pendulum, the *acceleration* of the object (which is the second differential) is equal to the *displacement* of the object but in the *reverse* direction. The result is always a sinusoidal oscillation.

The hyperbolic functions also have interesting gradient functions. The gradient of the $\cosh x$ function is $\sinh x$, and the gradient of the $\sinh x$ function is $\cosh x$ (these two functions are unique in this respect). Differentiating *twice* gets you back to the original function (*opposite*).

Although it has a discontinuity at $x = 0$, the function $f(x) = \frac{1}{x}$ nevertheless has the derivative $f'(x) = -\frac{1}{x^2}$ (*lower, opposite*). Recall that $f(x) = \frac{1}{x}$ is itself the derivative of $\log_e x$ (*page 29*).

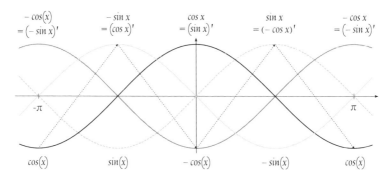

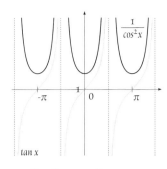

THE TANGENT GRADIENT

$$(\tan x)' = \frac{1}{\cos^2 x}$$

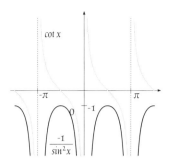

THE COTANGENT GRADIENT

$$(\cot x)' = \frac{-1}{\sin^2 x}$$

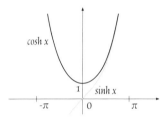

THE SINH GRADIENT

$$(\sinh x)' = \cosh x$$

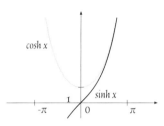

THE COSH GRADIENT

$$(\cosh x)' = \sinh x$$

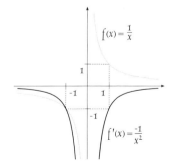

DERIVATIVE OF $^1/x$

$$f(x) = \frac{1}{x} \qquad f'(x) = \frac{-1}{x^2}$$

When x is very small $f(x)$ is very large. At $x = 1$, $f(x)$ has fallen to $+1$, and $f'(x)$ to -1. As x increases, the derivative gradually reduces to zero. Like the tangent and cotangent function above, although this function is discontinuous at $x = 0$ it can still be differentiated everywhere else (see p.28).

CALCULUS
the fundamental theorem

The top graph opposite shows a section of an arbitrary curve, $f(x)$, with two points A and B. Moving from A to B, x increases slightly, by δx (where δx, or 'delta x', means 'the change in x'). This causes the function $f(x)$ to increase slightly, by δy. If δx is small we can say that

$$\text{gradient at } x \approx \frac{\delta y}{\delta x} \qquad \text{and} \qquad \delta y \approx \delta x \times \text{gradient at } x$$

The second graph shows the gradient of the top graph, found by differentiation (*page 30*). The area added under this curve between A and B is approximately equal to $\delta x \times$ the gradient at x. But this is none other than δy. So while graph 2 plots the *gradient* of graph 1, graph 1 plots the *area under the line* of graph 2. What this means is that finding the area under a graph (*integration*) is the exact reverse of finding the gradient of a graph (*differentiation*). There is one important caveat. During the process of differentiation, any constants are rejected (because the gradient of a graph is unaffected by the addition of a constant). So when we reverse the process to find the area under a graph, we may have to add a constant back in. For example, to calculate the area under the line $y = 2x - 2$, we simply reverse the formula for each term we met on page 28, so that:

$$f(x) = ax^n \implies \int f(x) = \frac{a}{n+1} x^{n+1}$$
$$\text{and, } f(x) = 2x - 2 \implies \int f(x) = x^2 - 2x + C$$

where $\int f(x)$ is the *integral* of $f(x)$ and differentiating this integral produces the original function. In fact, you do not need a constant when integrating a polynomial, as here; you only need one when the integrand is not zero at $x = 0$ (e.g. the integrand of e^x is e^x, and $e^0 = 1$).

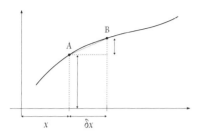

GRAPH 1: SOME CURVE $f(x)$.
A section of a typical curve. Note that at A the gradient of the curve is positive and then decreases as we move to B. Graph 1 plots the area under Graph 2. The approximations become more and more accurate as δx is made smaller and smaller.

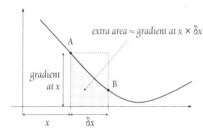

extra area ≈ gradient at $x \times \delta x$

GRAPH 2: GRADIENT OF GRAPH 1
A plot of the gradient of Graph 1. Note how at A the value is positive, and then decreases as we move to B. The area under this graph is plotted by Graph 1.

INTEGRATION
The function $f(x) = 2x - 2$ (black line); and its integral $\int f(x) = x^2 - 2x$ (grey parabola), which tells us the total area under the graph from the origin up to some value of x. No constant is needed because the integral is zero at $x = 0$. Between $x = 0$ and $x = 1$ the original function is negative, and the graph of the area function is also negative. At $x = 1$ the original function turns positive and the area function starts to increase. Only at $x = 2$ has the area function made up the lost ground (because the shaded area counts as negative area).

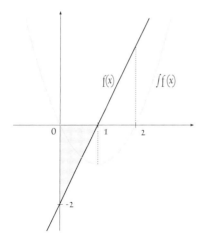

$f(x)$ $\int f(x)$

As a worked example, the area under the line between $x = 1$ and $x = 2$ is the area under the graph at $x = 2$ minus the area at $x = 1$:

$$\int_1^2 f(x) = \left((2)^2 - 2(2)\right) - \left((1)^2 - 2(1)\right)$$
$$= 0 - (-1) = 1$$

INTEGRATION
functions from functions

A good way to think of calculus is shown opposite. Suppose you drive from home to see a friend and every second record the speed *s* of your car and the total distance *d* travelled so far. While *s* measures the rate at which *d* is increasing (i.e. *s* is the gradient or differential of *d*), *d* is the sum of all the little distances travelled every second from the start until now (i.e. *d* is the integral of *s* since the start of the journey).

Not all functions can be differentiated (*see p. 28*). However, unless it jumps off to ∞ (e.g. $f(x) = 1/x$, at $x = 0$, *opposite*) you can nearly always calculate the area under a graph, finding its integral even if it has jumps or kinks in it, to give a continuous (but not always smooth) function.

Mathematicians sometimes find integrals by differentiating potential candidates until they stumble on the answer. For example, try to integrate the expression $y = 1/x$. On page 29 we saw that the gradient (differential) of the log function was $1/x$. It follows that the integral (area under the graph) of $1/x$ must be $\log_e(x)$ plus some constant (*e.g. see opposite, and p. 32*).

Sometimes it is not even possible to write down the integral of a simple function using standard algebraic and exponential functions. A good example is $\sin(x^2)$, whose differential and integrals are pictured (*right*).

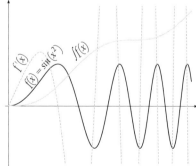

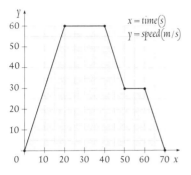

$x = time(s)$
$y = speed(m/s)$

A SHORT JOURNEY: Each square in the graph represents a distance travelled of $25\,m$ (i.e. $5\,m/s$ for $5s$). The total distance travelled is the total area under the graph — the integral of the speed.

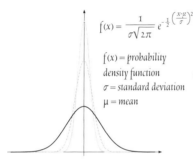

$$f(x) = \frac{1}{\sigma\sqrt{2\pi}}\, e^{-\frac{1}{2}\left(\frac{x-\mu}{\sigma}\right)^2}$$

$f(x) = $ probability density function
$\sigma = $ standard deviation
$\mu = $ mean

NORMAL DISTRIBUTION
The graphs opposite illustrate how different properties of a population of a million people cluster round the mean value. Some properties (like height) do not vary very widely so the curve is tall and thin; other properties (like annual income) vary greatly so the curve is more spread out. The area under each of these graphs, however, must be equal to 1 million.

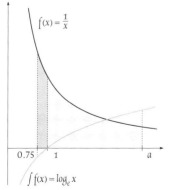

$f(x) = \dfrac{1}{x}$

$\int f(x) = \log_e x$

INTEGRATION OF $y = 1/x$
Since the gradient (differential) of the $\log_e$ function is $1/x$ (see p.29) it follows that the integral of $f(x) = 1/x$ must be the $\log_e$ function.
1. The area under the curve $1/x$ between $x = 1$ and $x = a$ is $\log_e(a) - \log_e(1) = \log_e(a)$.
2. The area under the curve between $x = 0.75$ and $x = 1$ is $\log_e(1) - \log_e(0.75) = 0.28768\ldots$
3. The area under the curve between $x = 0.75$ and $x = a$ is $\log_e(a) - \log_e(0.75)$
$= \log_e(a) + 0.28768\ldots$

THE GAMMA FUNCTION
and other interesting integrations

The factorial $n!$ of a positive integer n is the product of all the integers less than or equal to n, $1 \times 2 \times 3 \times \ldots \times n$. So a graph of $n!$ against n increases very rapidly, rather like an exponential curve, only faster.

The factorial function only works for the integers, and there is no simple function (involving only algebraic and exponential operations) which 'joins the dots'. However, we can specify its desired behaviour precisely, by first noting that $6! = 6 \times 5!$ and $20! = 20 \times 19!$ etc. This means that for any n, $n! = n \times (n-1)!$. Let us look for some function $\Gamma(n)$ (pronounced 'gamma n') which satisfies the relation

$\Gamma(n) = n \times \Gamma(n-1)$ for all n, even when n is not an integer.

It turns out that the total area under the graph (from 0 to ∞) of

$$f(x) = \frac{x^{(n-1)}}{e^x} \quad \text{has exactly this property.}$$

And when n is a whole number, the area under the graph $= n!$.

So: $\Gamma(n)$ = area under the graph from 0 to ∞ of $\dfrac{x^{(n-1)}}{e^x}$

The logarithmic integral "li" function (*right*) is the integral of the reciprocal of the logarithm function. Surprisingly, as x increases, $\text{li}(x)$ becomes a good approximation to the prime-counting function, $\pi(x)$, the number of prime numbers less than or equal to x.

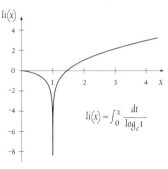

$$\text{li}(x) = \int_0^x \frac{dt}{\log_e t}$$

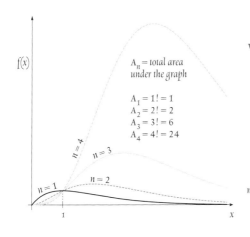

A_n = total area
under the graph

$A_1 = 1! = 1$
$A_2 = 2! = 2$
$A_3 = 3! = 6$
$A_4 = 4! = 24$

$n = 4$

$n = 3$

$n = 2$

$n = 1$

f(x)

1

x

THE GAMMA FUNCTION

When n is a whole number, the total
area under the graph of

$$f(x) = \frac{x^{n-1}}{e^x}$$

is equal to n!,
where n! is the factorial of n,

$$n! = 1 \times 2 \times 3 \times 4 \times \ldots \times n$$

The gamma function is used to
extend the idea of the factorial
function to include fractional,
negative and even complex numbers.
E.g. gamma (0.5) equals $\sqrt{\pi}$.

Once you have grasped the idea of a function as a mathematical entity in its own right, a whole new
world opens up in which you can study 'OPERATORS' which turn functions into other functions.
The most important of these are the operations of DIFFERENTIATION and INTEGRATION.

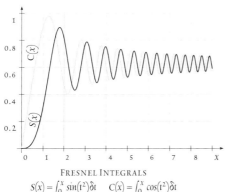

FRESNEL INTEGRALS

$$S(x) = \int_0^x \sin(t^2)\delta t \qquad C(x) = \int_0^x \cos(t^2)\delta t$$

The Fresnel integrals are transcendental functions,
like the exponential, logarithmic and trigonometrical
functions, i.e. they are not algebraic, nor polynomial.

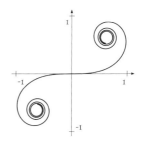

EULER SPIRAL

$$(x, y) = (C(L), S(L))$$

A parametric of the Fresnel integrals. The
curvature at any point is proportional to its
distance along the spiral from the origin.

SURFACE FUNCTIONS
into the third dimension

So far we have just considered functions of a single variable x (or, in the case of parametric functions, t). We now turn to functions of two variables x and y, which can be thought of as representing a point on a plane so that $f(x, y)$ can be thought of as the height above the plane. In other words, $f(x, y)$ defines an undulating surface.

Look in your pantry! The shape of an egg tray can be described by the function $f(x, y) = \sin x + \sin y$. And the hyperbolic paraboloid looks just like a saddle-shaped potato chip: $f(x, y) = x^2 - y^2$ (*see opposite*).

The famous Mandelbrot set is often depicted as a three-dimensional surface (*below*). A more regular three-dimensional surface is also shown (*below right*). Sometimes a function is better defined in terms of polar coordinates r and θ; e.g. the surface of a pond after a stone has been thrown in can be approximated by a polar function (*opposite, lower right*).

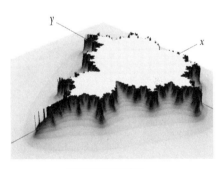

MANDELBROT SET
$z_{n+1} = z_n^2 + C$, where z and C are of the form $(ax + biy)$

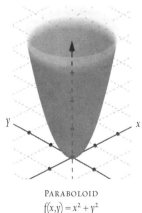

PARABOLOID
$f(x,y) = x^2 + y^2$

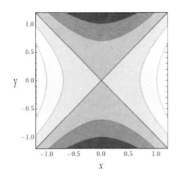

Saddle Chip — Hyperbolic paraboloid

$$f(x,y) = x^2 - y^2$$

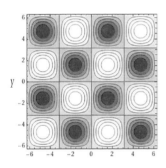

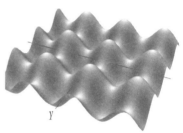

Egg Box — Sine wave surface

$$f(x) = sin(x) + sin(y)$$

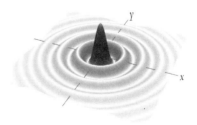

Water Ripples — Cosine surface (polar)

$$f(r,\theta) = \frac{cos^2(5r)}{e^{r^2}}$$

Vector Functions
fields they create

Another class of functions takes as its input the points in the plane and outputs a *vector*, a quantity **a** with a magnitude $\|\mathbf{a}\|$ and a *direction*. E.g. the vector $\mathbf{a} = (3, 4)$ has magnitude 5 and direction $53.1°$ (*below*).

Vectors can be visualised by using 'field lines' and 'weather' arrows whose length and thickness indicate the strength of the field (*see opposite*) expressed in either Cartesian or polar coordinates (*below*). It is often simpler to use the latter, where for every point in the plane (x, y) the vector function specifies a magnitude (r) and a direction θ. E.g., the electric field at (x, y) round a positive point charge placed at the origin has

$$\text{magnitude } r = \frac{k}{x^2 + y^2} \quad \text{and} \quad \text{direction } \theta = \arctan\left(\frac{y}{x}\right)$$

(where k is a constant depending on the strength of the charge; note that this field obeys an inverse square law). Likewise, the magnetic field at (x, y) around a current-carrying wire has

$$\text{magnitude } r = \frac{k}{\sqrt{x^2 + y^2}} \quad \text{and} \quad \text{direction } \theta = \arctan\left(\frac{x}{y}\right)$$

(This field is inversely proportional to the distance of the point from the wire, with the direction of the field at right angles to the radial line.)

To specify more complex fields, like that of a bar magnet or the flow of air over an aerofoil, simply add the component fields together using vector addition (*right*).

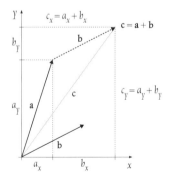

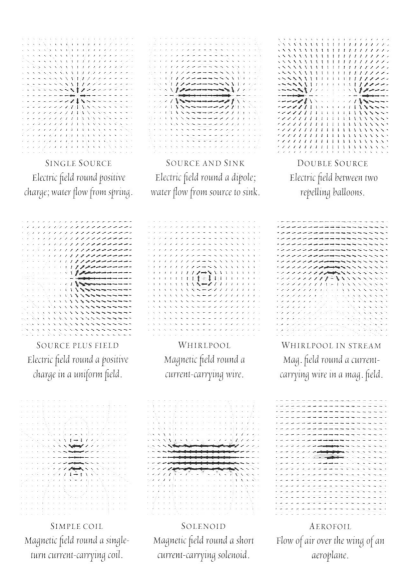

SINGLE SOURCE
Electric field round positive
charge; water flow from spring.

SOURCE AND SINK
Electric field round a dipole;
water flow from source to sink.

DOUBLE SOURCE
Electric field between two
repelling balloons.

SOURCE PLUS FIELD
Electric field round a positive
charge in a uniform field.

WHIRLPOOL
Magnetic field round a
current-carrying wire.

WHIRLPOOL IN STREAM
Mag. field round a current-
carrying wire in a mag. field.

SIMPLE COIL
Magnetic field round a single-
turn current-carrying coil.

SOLENOID
Magnetic field round a short
current-carrying solenoid.

AEROFOIL
Flow of air over the wing of an
aeroplane.

MAPPING FUNCTIONS
projecting & distorting our worldview

Mapping functions take a point on the plane (x, y), or (r, θ), and return another point on the plane. Restricting ourselves to linear functions, the general form of such a mapping (called an *affine transform*) is

$$x'(x, y) = ax + by + e \qquad y'(x, y) = cx + dy + f$$

Using appropriate values of a, b, c, d, e and f we can translate (move sideways), rotate, magnify and shear the plane in any combination. Ignoring the translation effect of e and f, we can write these transformations in the form of a 2×2 matrix of a, b, c and d:

$$\begin{bmatrix} x' \\ y' \end{bmatrix} = \begin{bmatrix} a & b \\ c & d \end{bmatrix} \begin{bmatrix} x \\ y \end{bmatrix} = \begin{bmatrix} ax + by \\ cx + dy \end{bmatrix}$$

Examples are shown (*opposite*). Successive transformations can be achieved by multiplying the matrices together using matrix algebra.

If we allow the use of non-linear functions, then the possibilities become enormous. For example, to map points on a globe (with latitude ϕ and longitude λ) onto a flat sheet of paper (x, y), the Mercator projection uses the function:

$$(\phi, \lambda) \implies x = k\theta; \; y = \frac{k(1 + \sin\theta)}{\cos\theta}$$

where the scale factor $k = \sec\phi$. Although very distorted in high latitudes, its advantage was that sailors could draw a straight line on the map and follow it using a constant compass bearing.

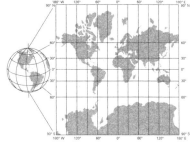

TRANSLATE

$$x' = x + e$$
$$y' = y + f$$

by e (along x) and f (along y)

SHEAR

$$\begin{bmatrix} x' \\ y' \end{bmatrix} = \begin{bmatrix} 1 & k \\ 0 & 1 \end{bmatrix} \begin{bmatrix} x \\ y \end{bmatrix}$$

where k is the shear factor

ROTATE

$$\begin{bmatrix} x' \\ y' \end{bmatrix} = \begin{bmatrix} \cos\theta & \sin\theta \\ -\sin\theta & \cos\theta \end{bmatrix} \begin{bmatrix} x \\ y \end{bmatrix}$$

where θ is the angle of rotation

SCALE

$$\begin{bmatrix} x' \\ y' \end{bmatrix} = \begin{bmatrix} m & 0 \\ 0 & m \end{bmatrix} \begin{bmatrix} x \\ y \end{bmatrix}$$

where m is the magnification factor

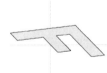

SHEAR + ROTATE

$$\begin{bmatrix} x' \\ y' \end{bmatrix} = \begin{bmatrix} \cos\theta & \sin\theta \\ -\sin\theta & \cos\theta \end{bmatrix} \begin{bmatrix} 1 & k \\ 0 & 1 \end{bmatrix} \begin{bmatrix} x \\ y \end{bmatrix}$$

ROTATE + SHEAR

$$\begin{bmatrix} x' \\ y' \end{bmatrix} = \begin{bmatrix} 1 & k \\ 0 & 1 \end{bmatrix} \begin{bmatrix} \cos\theta & \sin\theta \\ -\sin\theta & \cos\theta \end{bmatrix} \begin{bmatrix} x \\ y \end{bmatrix}$$

ABOVE: Linear affine transforms, with their associated matrices. Note that the effect of a shear followed by a rotation is not the same as the effect of a rotation followed by a shear (non-commutative). In contrast to most non-linear transforms, the inverse of a linear transform is also a linear transform. This means there is a one-to-one correspondence between points in the original and the transformed planes, and all linear transforms transform straight lines into straight lines.

COMPOSITE FUNCTIONS
function of a function

Sometimes the output of a function can involve other functions, either as a sum (or similar), e.g. $f(x) = f(b) + f(c) - f(d)$ or as a *composite function*, e.g. $f(g(h(x)))$. Consider the map application on your smartphone. There are many ways of picking a route from A to B—by time, distance, fuel economy, etc. For each route the application may calculate a overall parameter, but this parameter will be function of the chosen route, itself a function of another function of the initial input numbers.

Many physical systems try to maximise or minimise a quantity: light always tries to minimise its time of travel between A and B; soap film always tries to minimise its surface area (*see opposite*). One famous problem (solved by Christiaan Huygens in 1673) was to find the curve down which a ball would roll so that, whatever point it was launched from, it would always arrive at the bottom in the same time. The time needed for the ball to descend is a function of the shape of the slope, i.e. a function of a function. The answer is a *tautochrone* (an inverted cycloid, *below, and see page 18*). The problem was important because a solution was needed to build precise pendulum clocks.

For the cycloid $x = r(\theta - \sin\theta)$ $y = r(1 - \cos\theta)$
the time t of descent of any ball from any height is $\pi\sqrt{r/g}$,
where r is the radius of the cycloid circle
and $g = 9.81$ m/s².

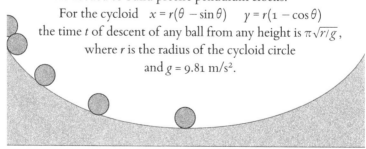

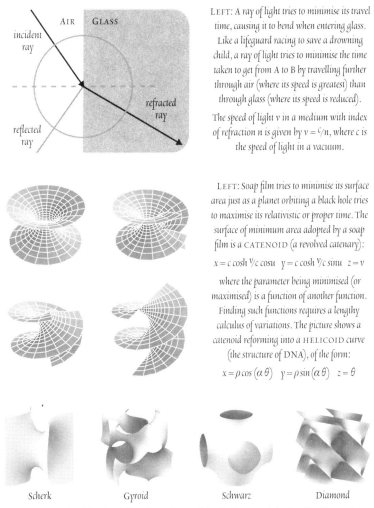

LEFT: A ray of light tries to minimise its travel time, causing it to bend when entering glass. Like a lifeguard racing to save a drowning child, a ray of light tries to minimise the time taken to get from A to B by travelling further through air (where its speed is greatest) than through glass (where its speed is reduced).

The speed of light v in a medium with index of refraction n is given by $v = c/n$, where c is the speed of light in a vacuum.

LEFT: Soap film tries to minimise its surface area just as a planet orbiting a black hole tries to maximise its relativistic or proper time. The surface of minimum area adopted by a soap film is a CATENOID (a revolved catenary):

$$x = c\cosh\tfrac{v}{c}\cos u \quad y = c\cosh\tfrac{v}{c}\sin u \quad z = v$$

where the parameter being minimised (or maximised) is a function of another function. Finding such functions requires a lengthy calculus of variations. The picture shows a catenoid reforming into a HELICOID curve (the structure of DNA), of the form:

$$x = \rho\cos(\alpha\theta) \quad y = \rho\sin(\alpha\theta) \quad z = \theta$$

Scherk   Gyroid   Schwarz   Diamond

ABOVE: Triply minimal periodic surfaces. The gyroid is a fascinating minimal surface discovered by Alan Schoen in 1970. It is defined by the equation $\cos x \sin y + \cos y \sin z + \cos z \sin x = 0$.

COMPLEX FUNCTIONS
real and imaginary

What is the square root of a negative number? It can't be a positive or negative number, since squaring either produces a positive number. One of the most remarkable discoveries ever made in mathematics is the discovery that the real numbers can be extended by turning the number line into a plane (*see below*) and introducing a new number i, the square root of -1. We now describe a *complex number*, such as $2 + 3i$, and treat it as if it was a perfectly ordinary number. However, instead of residing along a line, it can be represented by a point (x, y) in the Cartesian plane, or with polar coordinates, modulus r and argument θ.

Complex numbers can be operated on just like real numbers and we can consider functions of a complex number $f(z)$ in the same way that we can consider functions of a real number $f(x)$. The only problem is that the input of such a function has two variables (the 'real' and so-called 'imaginary' parts) and the output has two variables too.

To visualise the behaviour of a function like $f(z) = z^2$ we need a piece of four-dimensional graph paper! The best we can do on a two-dimensional page is to draw a perspective picture of the complex plane and just one aspect of the function (*see examples opposite*).

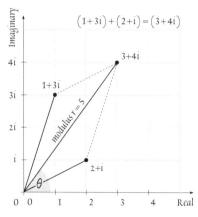

46

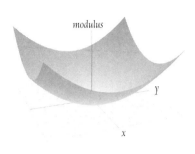

The MODULUS of z^2 only depends on the modulus of z so it is rotationally symmetric.

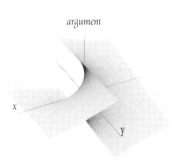

The ARGUMENT of z^2 is a double helix because as z rotates once about the origin, z^2 rotates twice.

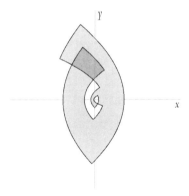

Another helpful and interesting way to think of a complex function such as z^2 is as a mapping function (see page 42) which maps the point (x, y) onto the point $(x^2 - y^2, 2xy)$. This is because $(x + iy)^2 = x^2 - y^2 + 2ixy$. The illustration shows the effect of this z^2 mapping on the letter F drawn in the complex plane, which is then distorted and wrapped round so that part of it overlaps itself.

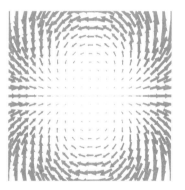

Yet another way is to think of the function as generating a vector at every point on the XY plane (see page 40) whose magnitude is the modulus of the result and which points in the direction of the argument. In the case of the function z^2, the vectors get larger as you get further from the origin and it is interesting to note that the vectors rotate round twice for every rotation round the unit circle.

THE COMPLEX EXPONENTIAL
and a remarkable identity

Perhaps the most important function of a complex variable is the complex exponential function:

$$f(z) = e^z = e^{x+iy} = e^x e^{iy}$$

To make sense of it, we need to interpret the expression e^{iy}. One way to do this is to interpret y as an *angle of rotation* (in radians, where 2π radians = $360°$). This means that the point in the complex plane represented by e^{iy} has Cartesian coordinates $(\cos y, \sin y)$. If $y = \pi$, then $e^{i\pi}$ represents the point one unit away from the origin but rotated by π (= $180°$), which is -1. And so we arrive at one of the most important equations in all of mathematics, Euler's identity: $e^{i\pi} = -1$.

More generally, we have the important result:

$$f(z) = e^{x+iy} = e^x(\cos y + i\sin y)$$

The plots opposite attempt to portray this. Since the modulus of e^{iy} is always 1 regardless of the value of y, the surface is just a flat sheet bent up into an exponential curve (*see opposite*). Plots of the real and imaginary parts of the function show the familiar exponential curve along the x-axis but displaying an oscillation in the y-direction, because y is a rotation and the function repeats every 2π radians.

Since multiplying any number by i is the equivalent of a $90°$ counterclockwise rotation in the complex plane, many trigonometric problems can be more easily solved by replacing the trigonometric terms with complex exponentials. This technique is widely used in engineering, electronics and Fourier analysis (*see page 26*). Often, the real and imaginary parts of the solution solve different problems!

The MODULUS of e^z is a kind of exponential ski-jump. It is not dependent on the y-coordinate.

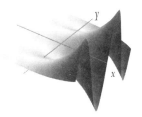

The ARGUMENT of e^z is just an inclined flat plane. It is not dependent on the x-coordinate.

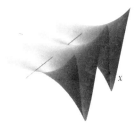

A slice through the REAL surface of e^z parallel to the x-axis is an exponential curve; parallel to the y-axis is a cosine curve.

A slice through the IMAGINARY surface of e^z parallel to the x-axis is an exponential curve; parallel to the y-axis it is a sine curve.

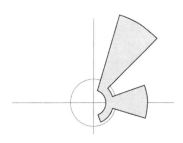

In the transformation of the letter F, horizontal lines are turned into radial ones and the vertical lines are turned into arcs of circles.

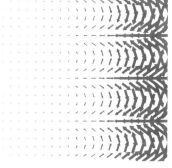

A vector map of e^z.

49

THE COMPLEX LOGARITHM
funnel and coil

Next in importance to the complex exponential function is the complex logarithm function $f(z) = \log z = \log(re^{i\theta}) = \log r + i\theta$ where $r = \sqrt{x^2 + y^2}$ and $\theta = \arctan(y/x)$ (i.e. the angle whose tangent is y/x).

The real component of $\log z$ is the logarithm of the modulus of z ($\log r$), and looks like a trumpet-shaped funnel which dips down to $-\infty$ at the origin (*opposite top left, and below left*). The imaginary part of $\log_e z$ is the argument of z (θ), and looks like a spiral helix (*opposite top right with one turn shown, and bottom right*), so for any given value of z (i.e. any point in the complex plane) there are infinitely many possible values of $\log_e z$.

The modulus of $\log z$ equals $\sqrt{(\log r)^2 + \theta^2}$. A careful look at the real and imaginary components of $\log z$ will reveal that the only place where both are zero is at the point $(1, 0)$. This is therefore the only point at which the modulus of $\log z$ is zero. At the origin the modulus of $\log z$ is infinite because the real component dips down to $-\infty$ at this point (*opposite centre left*). It should also be borne in mind that since the imaginary component has multiple values, the modulus (and the argument) of $\log z$ will also be multivalued.

The argument of $\log z$ is a strange mixture of a helix and an exponential curve (*opposite centre right*).

The REAL component of $\log(z) = \log(r)$.

The IMAGINARY component of $\log(z) = \theta$.

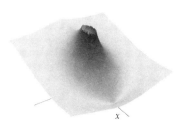

The MODULUS of $\log(z)$ is a kind of volcano mountain. It dips down to zero at the point $(1, 0)$.

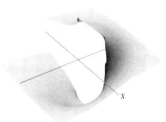

The ARGUMENT of $\log(z)$ is torn apart.

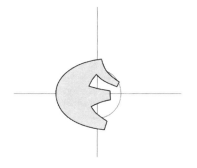

A transformed letter F.

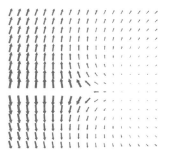

A vector map of $\log(z)$.

Complex Trigonometry
waves and spikes in four dimensions

We can now revisit the sine and cosine functions and look at them from a different perspective. Instead of expressing the exponential function in terms of the sine and cosine functions (*see page 48*), we can see that the (complex) exponential function is actually the more fundamental and express the latter in terms of the former. We start with:

$$e^{i\gamma} = \cos\gamma + i\sin\gamma \qquad\qquad e^{-i\gamma} = \cos-\gamma + i\sin-\gamma$$

Now since $\cos-\gamma = \cos\gamma$, and $\sin-\gamma = -\sin\gamma$, by adding or subtracting the two equations we find that:

$$\cos\gamma = \frac{e^{i\gamma}+e^{-i\gamma}}{2} \qquad\qquad \sin\gamma = \frac{e^{i\gamma}-e^{-i\gamma}}{2i}$$

Despite these equations involving i, they both evaluate to real numbers. So far, we have assumed that γ was a real number, but we can use the same equations to define the sine and cosine of a complex number z:

$$\cos z = \frac{e^{iz}+e^{-iz}}{2} \qquad\qquad \sin z = \frac{e^{iz}-e^{-iz}}{2i}$$

The upper four illustrations (*opposite*) depict the complex sine function (the complex cosine function is the same with a $\pi/2$ phase difference). The familiar sine and cosine curves appear along the real axis, but along the imaginary axis the hyperbolic functions sinh and cosh appear.

Also shown (*right*) are the modulus and vector representations of $\tan z$ (= $\frac{\sin z}{\cos z}$).

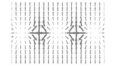

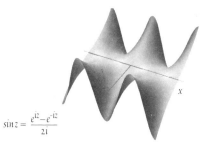

$$\sin z = \frac{e^{iz} - e^{-iz}}{2i}$$

The REAL component of $\sin z$, where z is a complex number. A sinusoidal oscillation along the (real) x-axis which increases in amplitude exponentially along the (imaginary) y-axis. The U-shaped curve along the y-axis is a cosh curve.

The IMAGINARY component of $\sin z$. This differs from the real component in that the oscillations on either side of the x-axis are out of phase, so peaks now face troughs. The curve along the y-axis is, of course, a sinh curve.

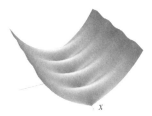

The modulus of $\sin z$.

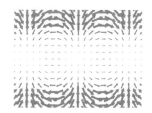

The complex function $\sin z$ in vector format.

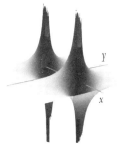

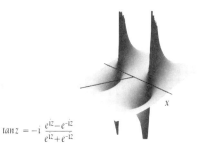

$$\tan z = -i\,\frac{e^{iz} - e^{-iz}}{e^{iz} + e^{-iz}}$$

The real component of $\tan z$.

The imaginary component of $\tan z$.

THE RIEMANN FUNCTION
crowned with the prime

The sum of the integer reciprocals, $1/1 + 1/2 + 1/3 + 1/4 + \ldots$ (the 'harmonic series') is infinite, whereas the sum of their squares, i.e. $1/1^2 + 1/2^2 + 1/3^2 + 1/4^2 + \ldots$, is finite, and (as Euler proved with some difficulty in 1735), equal to $\pi^2/6$. Euler's *zeta function*, where $s > 1$, is:

$$\zeta(s) = 1/1^s + 1/2^s + 1/3^s + 1/4^s + \ldots$$

While it is easy to calculate an approximate value for $\zeta(s)$, it is very difficult to calculate an exact value. For example, $\zeta(3) = 1.202056903\ldots$, but nobody knows whether this is related to π or to any other constant.

Euler did prove that the zeta function has a deep and fundamental relationship with the sequence of prime numbers $2, 3, 5, 7, 11, \ldots$, and in the early years of the twentieth century the zeta function was extended by replacing s with a complex number $z (= x + iy)$.

A plot of the modulus (magnitude) of the complex zeta function consists of a ridged plane which shoots up to infinity at the point $z = 1$ (the familiar harmonic series). Elsewhere, the function has a finite modulus, except in a few places where it dips down to zero.

The first zero occurs at the point $z = 0.5000000\ldots + 14.134725\ldots i$, (*see opposite*). Billions more zeros have been calculated and (apart from a series which lie to the left of the imaginary axis) all of them have a real part exactly equal to 0.5. There must be some reason why all the zeros lie on the line $z = 0.5 + iy$, but nobody has yet been able to prove that they do.

This is called the Riemann hypothesis and there is eternal fame and a prize of a million dollars waiting for the first person to prove it!

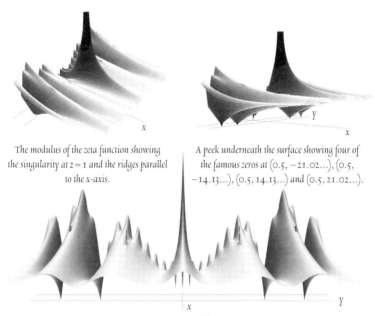

The modulus of the zeta function showing the singularity at $z = 1$ and the ridges parallel to the x-axis.

A peek underneath the surface showing four of the famous zeros at $(0.5, -21.02...)$, $(0.5, -14.13...)$, $(0.5, 14.13...)$ and $(0.5, 21.02...)$.

The modulus of the Riemann zeta function in the region $\zeta(z) > 0$. A view down the x-axis giving a clear view of the four zeros. The line $\zeta(z) = 1/2$ is indicated (the vertical dimension has been greatly reduced).

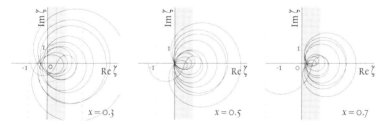

ABOVE: Images showing how $\zeta(z)$ wanders around as z moves along the lines whose real component x is 0.3, 0.5 and 0.7. When x is less than 0.5, z circles around the origin; when x is greater than 0.5, z stays outside of the origin; only when x is exactly equal to 0.5 does the line of its wanderings pass right through the origin. To prove the Riemann hypothesis, all you have to do is to prove that when $x > 0.5$, z never circles the origin, but that when $x < 0.5$ it always does.

THE SCHRÖDINGER EQUATION
all of chemistry in a box

At the heart of modern chemistry is Schrödinger's time-independent wave equation. It can take many forms, but in the form shown below it describes how a particle (such as an electron) of mass m and energy E behaves when confined to a region of space by a potential field V_p (e.g. within an atom or molecule).

$$\nabla^2 \Psi_p = \frac{-2m}{\hbar^2} \left(E - V_p\right) \Psi_p$$

Ψ_p, like V_p, is a variable which has different values at different points in space (hence the subscript p); it is a complex number and the square root of its modulus is the probability of finding the electron at that particular point. The nabla symbol ∇^2 is the Laplace operator, and $\nabla^2 \Psi_p$ tells us how quickly Ψ changes from place to place. V_p describes the shape of the potential well in which the particle is sitting. $\hbar$ is the reduced Planck's constant. Thus, the rate at which Ψ_p changes from place to place is proportional to the excess energy $(E - V_p)$ of the particle multiplied by Ψ_p and in the opposite direction (the minus sign).

To predict how the electrons in the outer shell of a molecule will behave (and in the process predict the physical and chemical properties of a new wonder drug), simply replace V_p with a function which describes the shape of the potential well round the molecule and 'solve' the equation (usually using a massively powerful computer) as a function which specifies the value of Ψ_p around the molecule. Even the single electron in the simplest atom, hydrogen, has a whole series of possible 'solutions', each with a characteristic energy (*see opposite*).

Mathematical functions are at the heart of everything around us!

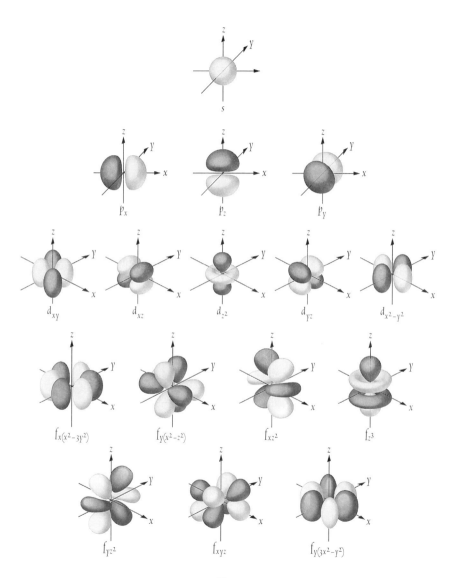

57

APPENDIX
functional operators

Functional operators take as their argument a FUNCTION rather than a quantity. As we have seen, functions can take on a wide variety of forms. In this book, where appropriate we use the expression $f(\,)$ to indicate a general function.

lim The LIMIT OPERATOR: e.g. $\lim_{x \to a} f(x)$ indicates the limit of the function as the specified variable x moves towards a. For example, $\lim_{x \to \infty}\left(\frac{1}{x}\right)$ is equal to zero because the graph of $1/x$ approaches the x-axis closer and closer as x increases.

$\sum$ The SUM OPERATOR: e.g. $\sum_{x=a}^{b} f(x)$ specifies the sum of a series in which x takes each integer value from a to b (a and b being integers). For example $\sum_{x=1}^{\infty}\left(\frac{1}{x}\right)$ indicates the sum $\frac{1}{1} + \frac{1}{2} + \frac{1}{3} + \frac{1}{4} + ...$ which is infinite.

$\prod$ The PRODUCT OPERATOR: $\prod_{x=a}^{b} f(x)$ is much less common but it works in exactly the same way as the sum operator except that the values of $f(x)$ are multiplied instead of summed. The most famous example is EULER'S PRODUCT formula: $\prod_{p \text{ prime}}\left(\frac{1}{1-p^{-s}}\right)$ which relates the Riemann zeta function to the product of an infinite series of functions involving all the primes.

$\int_{x=a}^{b}$ The DEFINITE INTEGRAL OPERATOR $\int_{x=a}^{b} f(x)\,dx$ is equal to the area under the graph of the function between the limits a and b (where a and b are not necessarily

integers). It is called a definite integral because the limits are specified.

$\int$ The INDEFINITE INTEGRAL OPERATOR $\int f(x)\,dx$ specifies a function whose gradient is equal to $f(x)$ and which, with the addition of an appropriate constant, can represent the total area under the graph between x and the origin.

$\frac{d}{dx}$ The DIFFERENTIAL OPERATOR $\frac{d\,f(x)}{dx}$ specifies a function which is equal to the gradient of the graph of $f(x)$ at every point x.

$\frac{\partial}{\partial x}$ The PARTIAL DIFFERENTIAL OPERATOR $\frac{\partial f(x)}{\partial x}$ specifies the gradient in the x direction if $f(\,)$ is a function of more than one variable (i.e. the ∂_x gradient with all the other variables held constant).

∇^2 The LAPLACE OPERATOR, OR LAPLACIAN is a second-order differential operator that has widespread usage in physics and engineering. It is a measure of the divergence of the gradient of a function. In the expression $\nabla^2 f = \nabla . (\nabla f)$, ∇^2 is the Laplace operator, ∇f is the gradient of the function, and $\nabla.$ represents the divergence.